Please leave your feedback, thoughts, and reviews on Amazon.

See Lt. Colonel McBrearty on CBS Los Angeles *Veterans' Voices.*

See Lt. Colonel McBrearty on CBS Los Angeles *Sports Central.*

Photo courtesy C.B.S., Los Angeles.

Follow John J. McBrearty's other literary works at:
JohnWritesHistory.com

AF882635

MILITARY SCHOOL

A Foundation for Success

Secrets For a Successful Life, Volume 2

by

John J. McBrearty

Lieutenant Colonel, U.S. Army (Retired)

Copyright © 2025 John J. McBrearty — All rights reserved. No part of this publication may be reproduced, stored in a retrieval system, or transmitted in any form or by any means, electronic, mechanical, photocopying, recording, or otherwise, without the author's prior written permission.
The views expressed in this publication are those of the author and do not necessarily reflect the official policy or position of Valley Forge Military Academy and College, Amazon, the Department of Defense, or the U.S. government.

ISBN: 978-1-966064-04-6 (Paperback)
ISBN: 978-1-966064-05-3 (Hardcover)

Library of Congress Control Number: 2025904192

The cover photo is courtesy of Valley Forge Military Academy and College. All uncredited photos in this book are the property of John J. McBrearty. Photos credited to other photographers are their property and appear as a courtesy in this book.
Published by *Military Books and More (MBM Publishing).*

Dedication

To my classmates of
Valley Forge Military Academy and College.

To my fellow veterans who stepped up and served our country.

To those veterans who made the ultimate sacrifice to be called Americans.

Attending a military school is not the easiest way to get an education.

However, I believe Chaucer got this concept right.

"Nothing ventured, nothing gained!"

I attended Valley Forge Military Academy and College for two years, and as a result, I became successful in life educationally, professionally, personally, and spiritually.

Thank you, Valley Forge.

John J. McBrearty

Warrior, Award-winning Author, Father
December 7, 2024

Foreword

Military School: A Foundation for Success; Secrets to a Successful Life, Volume 1, by John J. McBrearty, offers an engaging and deeply personal exploration of how a military school education can shape future success. As a retired Lieutenant Colonel in the U.S. Army, John draws upon his experiences from Valley Forge Military Academy and College to illustrate this structured educational path's profound and lasting impact.

What sets this book apart is its authenticity and practical wisdom. Through vivid storytelling, John reveals how the disciplined environment, emphasis on academic excellence, and rigorous leadership training at military school laid the groundwork for his remarkable career. His transformation from a young Sea Cadet to a distinguished Army officer is a testament to the lifelong benefits of this educational experience.

Particularly compelling are the chapters that showcase John's relentless drive to excel, including the prestigious gold star system and the invaluable lessons he carried into his professional life. His reflections not only narrate a personal success story but also guide prospective students and parents in considering military school as a pathway to discipline, leadership, and achievement.

Finally, John's book is an inspiring and enlightening read for anyone curious about the enduring value of a military school education. His journey proves that success is within reach with the proper foundation, discipline, and determination.

Dr. Derrick C. Darden, Ph.D.

CW3, U.S. Army (Retired)*; Academic* professor on Leadership, Organizational Behavior, & Organizational Management; author of *The Enemy in the Bush* and *Think Like a Leader*.

Contents

MILITARY SCHOOL

A Foundation for Success

Secrets For a Successful Life, Volume 2

Introduction

What did attending military school do for me?

By attending military school, I was not just afforded an education; the military lifestyle in which these military academies function takes things to an all-new level. These institutions not only provide a decent college preparatory education, but they also provide character development, leadership training, and a solid academic regiment. These virtues apply not only to future military leaders but also to up-and-coming corporate management. Traditional values are the core curriculum at a military institution. The structure provides an excellent environment for self-improvement and academic success. I attended Valley Forge Military Academy and College (V.F.), and most of the examples in this publication are from my first-hand experiences at "The Forge." However, I believe the other military schools identified in Appendix 1 also provide a similar experience to mine at V.F.

VFMAC Cadet John J. McBrearty, during his freshman year, 1978-79.
Photo courtesy of VFMAC.

Chapter 1

When Did You Decide That the Military Path Was Something You Wanted to Pursue?

My initial military exposure was through the U.S. Naval Sea Cadet Corps. I was a teenager, around 14 or 15 years old. From there, I found my niche and had a passion for the military. And I liked being in charge, the guy giving out orders to others. From my newfound calling, my grades in school improved, and so did my direction in life. At the end of high school, I knew I wanted to become a military officer, but my grades were not quite there. I knew I had to get a four-year college degree from a reputable university to get a commission in the military. I was not ready for the academic demands of such a program. Valley Forge Military Academy and College (V.F.) was my answer. They were a two-year, fully accredited junior college with an Army ROTC program. Graduating from this college would ensure my acceptance into a four-year university. So, I set my sights on V.F. with the thought in mind: study, study, and more study! Valley Forge helped me develop much-needed study habits that would last me a lifetime of learning.

Cadet John J. McBrearty (front left) with other students in a language laboratory in military school.
Photo courtesy of VFMAC.

Chapter 2

Were There Any Challenges During Your Time at Military School?

Attending military school was a very natural transition for me. As mentioned earlier, I already had military training with the Sea Cadets. I was also an athlete and always maxed the PT test with perfect scores. I just needed to work on my homework in the study hall environment. That's what I needed. I needed somebody to sit me down and make me crack the books. And that's what a military school does for you. They create an environment conducive to good study habits. The rest of my story is quite literally history, as you can read about in my various published books.

VFMAC Cadet John J. McBrearty during his Sophomore year, 1979-80.
Photo courtesy of VFMAC.

Chapter 3

What Did Your Time at Valley Forge Mean to You?

As they say, actions often speak louder than words. What did military school mean to me? I will demonstrate this by sharing some of my actions. As you can see, I have published many books on Amazon. In the Acknowledgment section of most of these books, there is a particular quote from me.

"To Valley Forge Military Academy and College, who gave me the academic foundation to become successful in life."

The moral of the story is represented by the Acknowledgments in these books, which are examples of me taking action. I not only say that military school helped me, but I put it in writing for the world to witness.

Cadet John J. McBrearty (far right) with other members of the Association of the U.S. Army. (Note: John was actually a U.S. Marine Corps Reserve member while attending VFMAC.) Photo courtesy of the VFMAC.

Chapter 4

What Aspects of Military School Helped You Thrive After Your Time at School?

What worked for me at Valley Forge was those gold stars. Valley Forge had a star system that represented your level of academic achievement. The red stars were C students, the silver stars were B students, and the gold stars were A students. I wanted those gold stars! I wanted those gold stars for three reasons: one, they looked cool; two, I wanted to be at the top of my class; and three, a cadet could earn additional on and off-campus privileges with gold stars.

I worked very hard to get those gold stars. They motivated me to achieve. That really worked for me, and it carried on for the rest of my life, influencing my career and my continued academics. I eventually earned a master's degree, and my family has doctorate degrees in medicine. So, attending military school greatly influenced not only my life but also those of my children and, hopefully, someday, my grandchildren as well.

2nd Lieutenant, John J. McBrearty, U.S.M.C., Amtrack Platoon Commander, 1983-85 timeframe, 29 Palms, CA.
Photo courtesy of Military Books and More Publishing.

Chapter 5

What Was Your Career Path Like After Military School?

I wrote an autobiography entitled *My Military Life*, but it hasn't been released yet. The sub-title for that book is *From Sea Cadet to Lieutenant Colonel United States Army, No Regrets*. I end the book by emphatically stating that I have no regrets about my career choices after serving 32 years in the military. I would do it all over again.

My autobiography goes into detail about my career path. To paraphrase, it goes something like this.

1. Following VFMAC, I attended Temple University and completed my B.A. degree. I participated in N.R.O.T.C. at Villanova University and was commissioned as an Officer of Marines.

2. I served on active duty and later in the reserves with the U.S.M.C.

3. During the 1990s, I was subject to the nation's R.I.F. (reduction in forces) and was released from the U.S.M.C. involuntarily. During my break of service between the U.S.M.C. and the U.S. Army, I briefly engaged in civilian business.

4. I affiliated with the Army National Guard and held various assignments of progressive responsibilities.

5. I participated in combat operations during Operation Iraqi Freedom.

6. When I retired from the A.G.R. (Active Guard Reserve) program, I had completed 32 years of military service.

7. Following my retirement from active duty, I participated in several Department of Veterans Affairs support groups.

8. I was introduced to the W.B.V. (Written by Veterans) writing support group. This connection resulted in the publication of my first book, *American History, a Veterans' Perspective*, which was based on my master's degree thesis.

9. I went on to publish many more books with the assistance of W.B.V. and several other writing support groups.

10. I wrote several books about my experiences with writing and publishing books in an effort to assist other veterans with their literary ambitions.

11. I co-created the Veterans and Law Enforcement Writing Support Group, composed of retired U.S. Army General Officers, several retired Police Chiefs, and a New York Times best-selling author.

12. I created Military Books and More Publishing, a service that assists veterans and their families with writing and publishing.

13. I conduct book signings, lectures, and speaking engagements about my successful life, career, and literary works.

During Operation Iraqi Freedom, John (left) attended the reopening of a local Iraqi school in the Babil Province of Iraq on May 25, 2004, following a joint civil affairs reconstruction project with the Polish Army. Photo courtesy MBM Publishing.

Chapter 6

How Have the Valley Forge Cornerstones Influenced Other Aspects of Your Life?

Again, referring to the premise, actions speak louder than words. I'll show you another example from one of my books. *Leadership Lessons from Operation Iraqi Freedom, Doing the Difficult Right and not the Easy Wrong*. This book came out of writing and publishing the *Combat Journal* book series. Many leadership challenges were identified in that day-to-day journal, particularly with the unit commander. You could call him "leadership challenged," to put it mildly. That leadership book practically wrote itself based on the actions of this "leadership-challenged" officer.

Upon that book's release, I was asked to speak on the topic of leadership at several universities. I then produced a lecture syllabus, which was eventually released as another book, *Army Values*.

Regarding the topic at hand, leadership, I always start and end my books with an original quote I created. I call them legacy quotes, and most of them originated from me. One such quote I borrowed from several notable people in history.

"Lead, follow, or get the Hell out of my way!"

Thomas Paine said that. George Patton said that, and Chesty Puller U.S.M.C. said that. Who else said that? Valley Forge Military Academy and College graduate Major John J. McBrearty, Operation Iraqi Freedom, said that. Again, actions speak louder than words.

On September 10, 2005, John participated in Hurricane Katrina relief operations in New Orleans, Louisiana. He was the Joint Task Force Operations Officer for the California National Guard. John's soldiers rescued and assisted numerous displaced civilians, landing these "weekend warriors" on Time Magazine's cover.
Photo courtesy MBM Publishing.

Chapter 7

What Advice Would You Give to Prospective VFMAC Students?

Choosing a school is a big decision. I suggest looking at the alumni of the various institutions. At Valley Forge, we have quite a few success stories within our alumni ranks. For example, H.R. McMaster was a classmate of mine, and we continue to correspond occasionally. Granted, he is higher up on the totem pole in the literary world, but, as the saying goes, he is good company to keep.

Other notable alumni include author J.D. Salinger; General Norman Schwarzkopf, former CENTCOM (United States Central Command) and Desert Storm Commander; General Gus Perna, former Commanding General of the U.S. Army's Material Command and Commander of Operation Warp Speed (initial vaccination distribution during the COVID-19 pandemic); and Pastor and author Dr. Lance Wallnau.

Secondly, I strongly recommend visiting several of these proposed schools and taking a campus tour. You will get a better idea of the environment and meet actual students, which will help you make an informed decision.

Attending military school was an excellent decision for me. I suggest putting your toe in the water first: try a semester and see if you like it. Then, try a second semester and reevaluate your year at military school. You can make the correct decision for yourself because you are making a well-thought-out and calculated decision.

This is how I approached the subject of attending military school or not. My parents frontloaded the first year's tuition. However, they clarified that any further college would be at Temple University, where Dad was a full professor. At Temple, I could attend for free.

You might wonder why I want to stay another year at Valley Forge. The answer was clear as a bell to me. I developed many fantastic relationships, my grades were quite good, and I ended up graduating second in my class. I was a Magna Cum Laude graduate and member of Phi Beta Kappa, both of which are on my resume to this very day.

So, take the chance, listen to testimonials, look at the alumni and their track record, and then decide for yourself.

Chapter 8

Who Were the Impactful People During Your Time at VFMAC?

I developed lifelong friendships while attending military school. One of my former roommates is Brendan O'Connor. Brendan served in the Army Special Forces and was awarded the Distinguished Service Cross. He is one of our generation's most highly decorated war veterans. Brendan is an example of a friend from Valley Forge for life; he was even my best man at my wedding.

Former college roommate, Sergeant Major Brendan O'Connor (pictured on the right), is seen here pinning on John's C.A.B. (Combat Action Badge) in front of a National Guard Armory. Photo courtesy MBM Publishing.

In January 2000, I participated in Yamasakura at Camp Itami, Japan, where I bumped into another V.F. classmate, Bruce Gomez. Bruce was also with the National Guard and was in the same exercise. Bruce and I have kept in touch since our chance meeting in Japan.

VFMAC classmates participated in the Yamasakura Multi-National Exercise in Camp Itami, Japan, in January 2000. Photo courtesy MBM Publishing.

Sometime in 2004, while serving in Operation Iraqi Freedom, I again crossed paths with a fellow Valley Forge alumnus, Joe Barsasius. He was in a Civil Affairs reserve unit and visiting our forward operating base. In recent years, Joe and I volunteered on the VFMAC Alumni Association Board of Directors.

VFMAC alumni Joe Bartasius and John McBrearty served in Operation Iraqi Freedom in the 2004 timeframe.
Photo courtesy MBM Publishing.

Those kinds of friendships are obtained from attending military school. I graduated some 40 years ago. I am still in contact with our alumni association, various school events, and even doing charity golf tournaments here in California with other V.F. alumni. By the way, my books are in the V.F. library and museum and are required reading for several classes. V.F. has been really good to me. I got a lot out of the education, and my family, soldiers, and Marines benefited from that.

Photo courtesy MBM Publishing.

Chapter 9

How Did You Get Into Writing?

When I was a cadet at Valley Forge, there was another cadet in my class, a battalion commander or some kind of big shot at the school; his name was Tom Balish. Tom's whole family, his father, uncles, and even his grandfather attended our school. He went on to have a successful Army career in the aviation field. Tom and I have stayed in touch a little bit over the years. The reason I bring up Tom is that he was the editor of the school newspaper. I wrote short stories about a new cadet and their experiences at a military school. Tom kept hitting me up, "I need something for the paper, I need something....do you have anything? Please! Give me something for the paper!" Sometimes, I would create these little stories at the last minute to keep Tom happy. Those stories are out there in the world, floating around somewhere. That is, in part, how I got started. I later wrote some articles for military publications when we deployed. That gets me up to creating *Combat Essays. Combat Essays* is a series of essays and letters I wrote about our unit going to combat. I sent them back to local newspapers, publications, and families.

These individual essays were published all over the place, so I made a book out of them. I participated in several Veterans Affairs support groups, and along the way, I met a mentor, Andreas Kossak. He was a former professor at California State University and a member of the Writers Guild of America. Andreas and the W.B.V. writing support group helped me with my first book. And that's where I really learned the whole craft of how to put a book together, editing, formatting, photos, maps, charts, and marketing. Following that, I knocked out two shorter books. But again, by perfecting my craft, I eventually got into more significant projects like Comb*at Journal,* which might be turned into a movie sometime soon.

I started another book series entitled *How to Get Published on Amaz*on. This three-part series takes the reader from the initial steps of the writing process to self-publishing, marketing, and refining the process. I concluded that series with many examples from my work.

I eventually created a children's book series entitled *Shih Tzu Adventures*, featuring *Claire the Magical Dog*. These stories have a profound moral message.

I currently have over a dozen books published on Amazon, and I acknowledge Valley Forge Military Academy and College in most of them. I thank them for helping me get to this point in life.

Some of my books are Amazon Best Sellers in their respective categories. The purpose behind my writing is for the soldiers and the Marines I served with. I dedicate every book to the five soldiers we lost in combat. I never want them to be forgotten. I write for those soldiers, the soldiers I deployed with, their families, and fellow veterans.

John with wife Lynette in 2024.
Photo courtesy of Nichole Rea.

Chapter 10

What Else Would You Like to Share With Prospective or Current Military School Students?

When speaking at a High School, college, or university, I sometimes like to share some of my fatherly advice with the students and faculty. As I mentioned, picking a school is a huge decision, but an even more significant decision is choosing your life partner. I will use me as an example. My wife and I met while attending Temple University in Philadelphia, Pennsylvania. We dated for two years and were engaged during the third year, and this year marked our 40th wedding anniversary. She recently retired after teaching elementary school for over 31 years. Both she and I returned to academia for our respective master's degrees, all the while still working full-time. I believe our home academic environment influenced our daughter to persevere and complete medical school to become a physician. Our son recently completed his biochemistry degree and E.M.T. certification. He is also pursuing graduate-level education in the medical field.

The moral of this story is that attending a military school transcended beyond me as an alumnus. It permeated our family, friends, and coworkers, so I am forever grateful.

John landed a speaking role and his Screen Actors Guild card when the Hollywood movie Taps *was filmed at VFMAC. Photo courtesy of MBM Publishing.*

A funny side note: If you watch the film *Taps*, you just might see me. Hollywood came to Pennsylvania and filmed the feature motion picture *Taps* at Valley Forge Military Academy and College. I don't know how, but I got into that movie. If you look at the scene in which Tom Cruise, Timothy Hutton, and Sean Penn observe cadets doing the trick drill scene, I am the guy leading the drill. I got lucky with that one.

Lt. Colonel John J. McBrearty, Centurions Battalion Commander, CAARNG.
The photo is courtesy of the Department of Defense and is in the public domain.

Chapter 11

What Final Thoughts About Attending a Military School Would You Like to Share?

The virtues that Valley Forge instills in its cadets are still with me to this day. A Valley Forge cadet will not lie, cheat, or steal, nor tolerate those who do. I took those virtues to heart and carried them with me into my military service and throughout my life, and they really paid dividends.

I talk a lot about character in my books. Your character is your actions; it is what you do when people are not around to watch. That demonstrates your character. The military school greatly influenced me to develop the proper character to become a good officer, leading by example always. I also set out to become a good father and humanitarian.

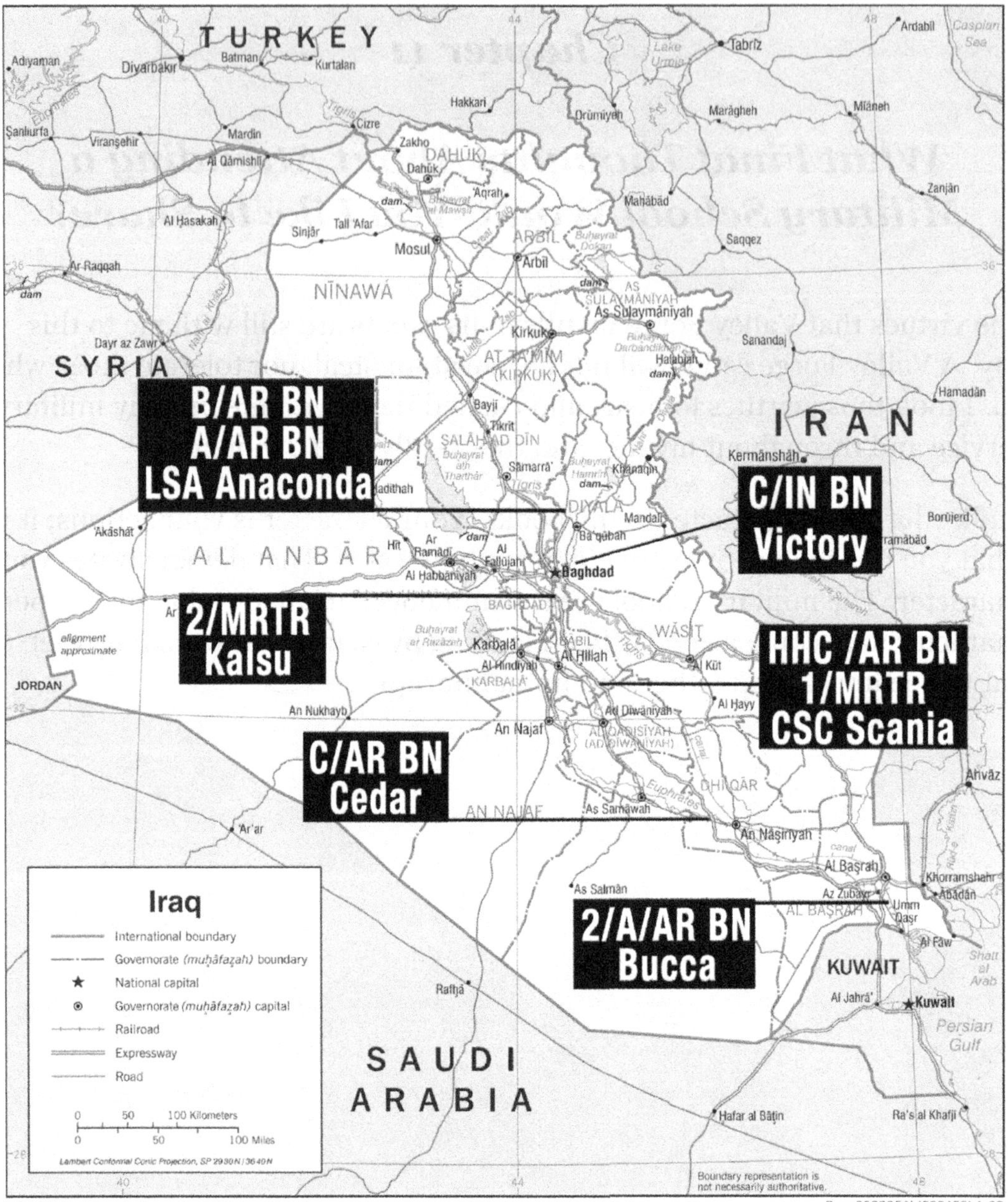

John's battalion was spread out over multiple F.O.B.s (Forward Operating Bases) during Operation Iraqi Freedom. John was second in command of these units.
Map by John J. McBrearty. Source: The World Factbook, CIA Maps.

Conclusion

You have witnessed a successful alumnus from a distinguished military school describe how the academy greatly influenced his future and set him on a trajectory in life filled with triumphs and successes. Perhaps this one example could be just the thing for the young man or woman on the fence about going to military school.

For John, attending military school was a slam-dunk decision. This decision resulted in his achieving a successful professional career while keeping his personal and spiritual growth at the forefront of his priorities.

Photo courtesy Nichole Rea

What is Next?

A VETERAN'S PATH TO SUCCESS

Secrets for a Successful Life, Volume 3

A Retired Military Officer Shares His Secrets for Having a Successful Life.

What is your definition of success?

By definition, success is often equated to obtaining a certain level of financial stability or professional achievement. In this manuscript, John discusses those achievements at length, but more importantly, the successful personal aspects of his life. John's philosophy incorporates the concept of being not only the physically successful individual but also the emotionally and spiritually successful person.

In this book, you will find John's simple yet effective life lessons, which, if followed, could also include you in this exclusive and prosperous club.

This book could quite literally change the trajectory of your life!

Military school cadets conducting a pass-in review.
Photo courtesy of the VFMAC.

Appendix 1
Military Schools in the U.S.A. and Canada

(In alphabetical order.)

Admiral Farragut Academy
St. Petersburg, FL
https://farragut.org

Army & Navy Academy
Carlsbad, CA
https://armyandnavyacademy.org

Benedictine College Preparatory
Richmond, VA
https://www.benedictinecollegeprep.org

Camden Military Academy
Camden, SC
https://camdenmilitary.com

Colorado Military Academy
Colorado Springs, CO
https://coloradomilitaryacademy.org

Culver Academies
Culver, IN
https://www.culver.org

Fishburne Military School
Waynesboro, VA
https://www.fishburne.org

Fork Union Military Academy
Fork Union, VA
https://www.forkunion.com

Georgia Military College,
Milledgeville, GA
https://www.gmc.edu

Hargrave Military Academy
Chatham, VA
https://hargrave.edu

Louisiana State University
Baton Rouge, LA
https://www.lsu.edu

Marine Military Academy
Harlingen, TX
https://www.mma-tx.org

Marion Military Institute
Marion, AL
https://marionmilitary.edu

Massanutten Military Academy
Woodstock, VA
https://www.militaryschool.com

Missouri Military Academy
Mexico, MO
https://www.missourimilitaryacademy.org

New Mexico Military Institute
Roswell, NM
https://www.nmmi.edu

Norwich University
Northfield, VT
https://www.norwich.edu

Oak Ridge Military Academy
Oak Ridge, NC
https://www.oakridgemilitary.com
Randolph-Macon Academy
Mendota Heights, MN
https://rma.edu

Riverside Prep
Gainesville, GA
https://www.riversideprep.org

Robert Land Academy
Wellandport, Ontario, CAN
https://robertlandacademy.com

Saint Thomas Academy
Mendota Heights, MN
https://www.cadets.com

Southern Preparatory Academy
Camp Hill, AL
https://www.southernprepacademy.org

St. John's Northwestern Academies
Delafield, WI
https://www.sjnacademies.org

Tarleton State University
Stephenville, TX
https://www.tarleton.edu/cadets

Texas A&M Maritime Academy
Galveston, TX
https://tamug.edu/corps/index.html

Texas A&M University
College Station, TX
https://tamu.edu

The Citadel
Charleston, SC
https://www.citadel.edu

University of North Georgia
Dahlonega, GA
https://www.ung.edu/military-college-admissions/index.php

Utah Military Academy
Riverdale, UT
https://www.utahmilitaryacademy.org

Valley Forge Military Academy
Wayne, PA
https://www.vfmac.edu

Valley Forge Military College
Wayne, PA
https://www.vfmac.edu

Virginia Military Institute
Lexington, VA
https://www.vmi.edu

Virginia Tech Corps of Cadets
Blacksburg, VA
https://vtcc.vt.edu

Virginia Women's Institute for Leadership at Mary Baldwin University
Staunton, VA
https://marybaldwin.edu

Source:
Association of Military Colleges & Schools of the United States (AMCSUS)
https://amcsus.org

Several of John's published books.

Appendix 2
Books by John J. McBrearty

<u>Leadership Books</u>

LEADERSHIP LESSONS,

Volume 1, *From Operation Iraqi Freedom*

Volume 2, *Army Values, Lessons in Leadership*

Volume 3, *OIF Leadership 101*

<u>American Military History Books</u>

AMERICAN HISTORY, a VETERAN'S PERSPECTIVE,

Volume ***I***, *Essays, Reflections, and Reviews*

Volume II, *Combat Essays*

COMBAT JOURNAL, Part 1, *Operation Iraqi Freedom*

COMBAT JOURNAL, Part 2, *Battle for Iraq*

COMBAT JOURNAL, Part 3, *A Soldier's Journey to Hell*

COMBAT JOURNAL, Part 4, *Coming Home*

COMBAT JOURNAL, Part 5, *War is Hell*

How to Write and Get Published Books

LITERARY WORKS OF JOHN J. MCBREARTY

HOW TO GET PUBLISHED ON AMAZON

Volume 1, *A Guide for Beginning Writers*

Volume 2, *From Publishing Zero to Publishing Hero!*

Volume 3, *Inspirational Quotations and Other Writings by a Combat Veteran*

Volume 4, *How to Write a Book Review*

Sports Books

CHILDREN'S GOLF (1), *Teach the Golf Swing in Four Easy Steps*

CHILDREN'S GOLF (2), *Driving Range 101*

CHILDREN'S GOLF (3), *Golf Course 101*

CHILDREN'S GOLF (4), *Advanced Golf Techniques Anyone Can Learn*

Children's Books

SHIH TZU ADVENTURES

(CLAIRE THE MAGICAL DOG SERIES)

CLAIRE THE MAGICAL DOG (1), *Claire Meets Dean and Jacob*

CLAIRE THE MAGICAL DOG (2), *Claire Introduces Baby to Golf*

CLAIRE THE MAGICAL DOG (3), *Claire Digs to China*

CLAIRE THE MAGICAL DOG (4), *Claire Celebrates Veterans Day*

CLAIRE THE MAGICAL DOG (5), *Claire and the Service Dog*

Motivational Books

SECRETS FOR A SUCCESSFUL LIFE

Volume 1, *Veterans' Survival Guide, PTSD is not a Four-Letter Word!*

Volume 2, *Military School, a Foundation for Success*

Volume 3, *A Veterans' Path to Success*

John is pictured here at a book signing during a local charity golf tournament.
Photo courtesy MBM Publishing.

Appendix 3

John J. McBrearty
Literary Affiliations

Written by Veterans, California State University

California Writers Club, Inland Empire

Military Writers Society of America

Diamond Valley Writers Guild

Temecula Valley Writers & Illustrators

Veterans Breakfast Club

Loma Linda Veterans Writing Workshop

Veteran and Law Enforcement Writers Support Group

Military Books and More Publishing *(MBM Publishing)*

Author John J. McBrearty.
Photo courtesy Nichole Rea

About the Author

John J. McBrearty

John J. McBrearty is a retired Lt. Colonel with over 32 years of military service. His career spanned the globe, with duty in Iraq (Operation Iraqi Freedom,) Kuwait, Japan, Australia, Mexico, and Thailand. He is a member of Phi Theta Kappa and a Magna Cum Laude graduate of the Valley Forge Military Academy and College (AA,) Temple University (BA,) and American Public University (MA.) John served as an Assistant Professor of Military Science at California State University and Claremont McKenna College.

John's other works have been published or featured in *Sports Illustrated, Stars and Stripes, Golf Week, Golf Digest, Callaway Magazine, Grizzly Magazine, Los Angeles Times, Orange County Register, Main Line Times, News of Delaware County, Daily Bulletin, San Bernardino Sun, Press-Enterprise, Veterans Breakfast Club Magazine, Military Writers Society of America Dispatches Magazine,* and others.

John J. McBrearty published over a dozen books at the time of this book's release, and that list is growing each month. He serves as a mentor and advisor for other veterans by participating in several Department of Veterans Affairs support groups. John also volunteers his time and effort to assist fellow veterans in their quest to write and learn *How to Get Published on Amazon.*

John founded Military Books and More Publishing (MBM Publishing), a literary service for veterans and their families that assists with the writing and publishing processes.

John won 1st Place in the National Veterans Creative Arts Festival for creative writing, which earned him a Veterans Creative Arts Scholarship.

Upon his retirement, John was recognized for excellence by President Barack Obama.

Special Thanks

I wish to thank my family: Lynette (wife), Kristina (daughter), John, Jr. (son), Dean, and Jacob (grandsons.)

I also want to thank all the soldiers and Marines I served with over the years, particularly the five soldiers my unit lost in combat.

To my literary mentors, Andreas Kossak, Valerie Ormond, Sue Rushford, Cindy Rinne, John Cole, Peter Schreiber, Owen McCue, Peter J. Gravett, and the other members of my writing support groups.

To the Will family: Nancy, Steve, Derek, and Chelsea, thank you for your loyalty and continued support.

To the Department of Veteran Affairs and their wonderful employees.

To my Lord and Savior.

God bless you all.

John J. McBrearty

Mr. Thomas Goldblum '69, '71C
former Director of Alumni Relations
Valley Forge Military Academy and College.
Photo courtesy of VFMAC.

In Memoriam

To Mr. Thomas Goldblum '69, '71C,
Director of Alumni Relations
Valley Forge Military Academy and College.

A dear friend, colleague, adjunct professor, and #1 alumnus.

Tom personified the character and traditions of this venerable institution. May you rest in peace, my brother.

Legacy

"One can judge their time here on earth by how many souls they inspired and influenced."

John J. McBrearty

December 7, 2024

Veterans
Crisis Line

Military
Crisis Line

1-800-273-8255 PRESS 1

Veterans
Crisis Line

NEW NUMBER,
SAME SUPPORT.

Dial 988 then Press 1.
Share it with your networks.

Made in the USA
Monee, IL
07 November 2026

67873712R20044